The Best VACATION Ever

by Stuart J. Murphy

illustrated by Nadine Bernard Westcott

HarperCollinsPublishers

To Lindsay and Nina—
who have given me
some of their best ideas ever
—S.J.M.

To Becky
—N.B.W.

HarperCollins®, 🐛®, and MathStart™ are trademarks of
HarperCollins Publishers Inc.

For more information about the MathStart series, please write to
HarperCollins Children's Books, 10 East 53rd Street, New York, NY 10022.

Bugs incorporated in the MathStart series design were painted by Jon Buller.

Library of Congress Cataloging-in-Publication Data
Murphy, Stuart J., date
  The best vacation ever / by Stuart J. Murphy ; illustrated by Nadine Bernard
Westcott.
     p.   cm.   (MathStart)
  "Level 2."
  Summary: A girl uses data-collecting and problem-solving skills to determine where
her family should go for a vacation.
  ISBN 0-06-026766-6. — ISBN 0-06-026767-4 (lib. bdg.)
  ISBN 0-06-446706-6 (pbk.)
  [1. Vacations—Fiction.   2. Problem solving—Fiction.   3. Stories in rhyme.]
I. Westcott, Nadine Bernard, ill.   II. Title.   III. Series.
PZ8.3.M935Vaf   1997   95-41191
[E]—dc20   CIP
                                                                    AC

2   3   4   5   6   7   8   9   10
                    ❖

The Best Vacation Ever

My family's very busy.
We always rush about.

Mom is coming in

6

when Dad is going out.

Charlie has his buddies.

Grandma never takes it slow.

9

We need a great vacation,

10

but we don't know where to go.

Maybe if I ask some questions

and write down the answers, too—

then we'll have some charts

that will tell us what to do.

Mom		
Dad		
Grandma		

Should we go somewhere warm?

Should we travel somewhere far?

What about excitement?

Could Fluffer come with us?

	warm	cool
Mom		×
Dad	×	
Grandma	×	
Charlie	×	
Me	×	
	④	1

	far	near
Mom		X
Dad		X
Grandma		X
Charlie	X	
Me	X	
	2	③

	fun	quiet
Mom		X
Dad		X
Grandma	X	
Charlie	X	
Me	X	
	③	2

	no	Fluffer
Mom		X
Dad	X	
Grandma		X
Charlie		X
Me		X
	1	④

Now I'll add my numbers
and see just what they show.

After looking at my charts,

I know exactly where to go!

warm
near
fun
Fluffer

To find the perfect place

wasn't really very hard.

Our best vacation ever is
right here in our backyard!

30

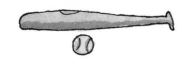

FOR ADULTS AND KIDS

I f you would like to have more fun with the math concepts presented in *The Best Vacation Ever*, here are a few suggestions:

- Read the story together and ask the child to describe what is going on in each picture. Talk about the questions the girl is asking and the answers that her family members give.

- Discuss what the girl learns from each of the charts she makes. Ask questions throughout the story, such as "Do more people want to go somewhere warm or cool? Do more people want to stay near or go far?"

- Ask the child to answer the girl's questions: "How about you? Would you rather take a vacation somewhere warm or cool? Somewhere exciting or quiet?"

- Together think of some questions that might help members of the child's family decide where to go on their own vacation. Write the questions down. Then help the child to ask the questions and record the answers in simple charts. Review the charts together and discuss what would be a good vacation destination.

- Look around your neighborhood and ask what things more people like. For example: What cars do more people drive—big or small? red or blue? What do more kids like to wear to school—sneakers or loafers? bright colors or light colors? Help the child to record this information and then discuss the answers to the questions.

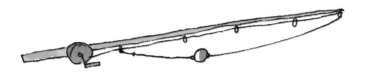

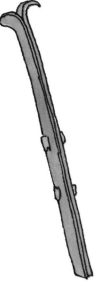
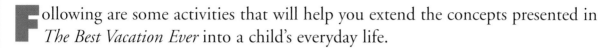

Following are some activities that will help you extend the concepts presented in *The Best Vacation Ever* into a child's everyday life.

Meals: Plan a picnic together. What questions would you ask to find out each person's favorite foods? How would you chart the information? Can you figure out a menu that most of the people would like?

Families: Together, make a family chart. Are there more females or males in your family? Are there more people with blue eyes or brown eyes? Do most people have the same color hair?

Favorite Days: Help the child make a chart that shows the days of the week across the top and the names of friends down the side. Have the child ask all the friends what day they like the most. Which days of the week are most often and least often the favorite? Why?

The following books include some of the same concepts that are presented in *The Best Vacation Ever*:

- WHAT MAKES ME HAPPY? by Catherine and Laurence Anholt

- THE CHOCOLATE CHIP COOKIE CONTEST by Barbara Douglass

- WHAT IS NEW? WHAT IS MISSING? WHAT IS DIFFERENT? by Patricia Ruben

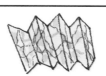